Made in the USA
Monee, IL
07 July 2026

56550376R00070